Future Energy

Solar Power

Julie Richards

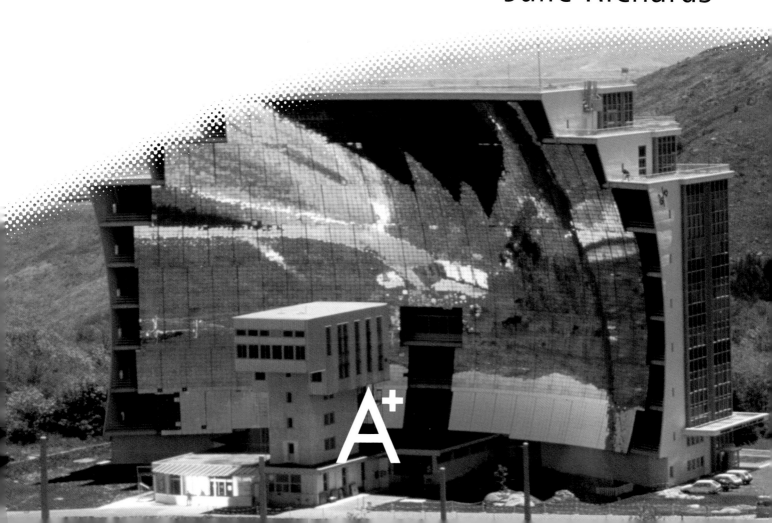

This edition first published in 2004 in the United States of America by
Smart Apple Media.

All rights reserved. No part of this book may be reproduced in any form or
by any means without written permission from the publisher.

Smart Apple Media
1980 Lookout Drive
North Mankato
Minnesota 56003

Library of Congress Cataloging-in-Publication Data
Richards, Julie.
  Solar power / Julie Richards.
  p. cm. — (Future energy)
  Contents: What is energy?—The sun as a source of energy—Where does the sun's energy come from?—Using solar energy—Solar energy through history—Early solar technology—Modern solar technology—Turning sunlight into electricity—Solar power at work—Solar power in the future —Advantages and disadvantages of solar power.
  ISBN 1-58340-332-9
  1. Solar energy—Juvenile literature. [1. Solar energy.] I. Title.
TJ810.3R52 2003
333.792'3—dc21   2002044639

First Edition
9 8 7 6 5 4 3 2 1

First published in 2003 by
MACMILLAN EDUCATION AUSTRALIA PTY LTD
627 Chapel Street, South Yarra, Australia 3141

Associated companies and representatives throughout the world.

Copyright © Julie Richards 2003

Edited by Anna Fern
Text and cover design by Cristina Neri, Canary Graphic Design
Illustrations by Nives Porcellato and Andy Craig
Photo research by Legend Images

Printed in Thailand

Acknowledgements
The author and the publisher are grateful to the following for permission to reproduce copyright material:

Cover photograph: Odeillo solar furnace, courtesy of Photolibrary.com.

AAP/AP Photo/Shizuo Kambayashi, p. 20; Argus—Fotoarchiv, pp. 13, 24; Robert Holmgren & Peter Arnold/Auscape International, p. 28; Richard Packwood—OSF/Auscape International, p. 11; T. Perrin—HoaQui/Auscape International, p. 7; Coo-ee Historical Picture Library, p. 9; Coo-ee Picture Library, pp. 8, 18 (top); Corbis Digital Stock, p. 16 (top); Energy Australia, p. 15; Dennis Sarson/Lochman Transparencies, p. 12 (top); Nasa, pp. 5 (bottom), 27, 29; Philips Lighting, p. 22; Photolibrary.com, pp. 1, 4, 10, 12 (bottom), 17 (top), 21, 23 (bottom), 30; Reuters, pp. 23 (top), 26; Solar Sailor/Paul Ruston, p. 25.

While every care has been taken to trace and acknowledge copyright, the publisher tenders their apologies for any accidental infringement where copyright has proved untraceable. Where the attempt has been unsuccessful, the publisher welcomes information that would redress the situation.

Contents

What is energy?	4
The Sun as a source of energy	5
Where does the Sun's energy come from?	6
Using solar energy	7
Solar energy through history	8
Early solar technology	10
Modern solar technology	11
Turning sunlight into electricity	12
Solar power at work	18
Solar power in the future	28
Advantages and disadvantages of solar power	30
Glossary	31
Index	32

Glossary words
When a word is printed in **bold** you can look up its meaning in the glossary on page 31.

What is energy?

Energy makes the world work. People, plants, and animals need energy to live and grow. Most of the world's machines are powered by energy that comes from burning coal, oil, and gas. Coal, oil, and gas are known as fossil fuels. Burning fossil fuels makes the air dirty. This is harmful to people and damages the environment.

Scientists are not sure how much longer fossil fuels will last. It depends on whether or not new sources of this type of energy are found, and how carefully we use what is left. Scientists do know that if we keep using fossil fuels as fast as we are now, they *will* run out. An energy source that can be used up is called non-renewable. A renewable source is one that never runs out. The world cannot rely on fossil fuels as a source of energy for everything. We need to find other sources of safe, clean, renewable energy to power the machines we have come to depend on.

These are the fossilized remains of a fish. Fossil fuels are the plants and animals that died millions of years ago and turned into coal, oil, and gas.

Solar Power

The Sun as a source of energy

All living things need energy to grow and stay healthy. Animals and people get their energy from food. The plants that animals and people eat get their energy from the Sun.

Sun

The plant changes sunlight into food.

Plant-eating animal eats the plant.

Meat-eating animal eats the plant-eater.

Plants store energy from the Sun. The Sun's energy is passed along the food chain giving energy to living things.

The Sun is a huge ball of hot, swirling gases. Energy from the Sun is called solar energy. Solar energy travels through space and reaches the Earth as light and heat. The Sun is a source of clean, safe, renewable energy. Every 45 minutes, the Earth receives enough energy from the Sun to power everything on our planet for a whole year.

The Sun is a huge ball of very hot gases.

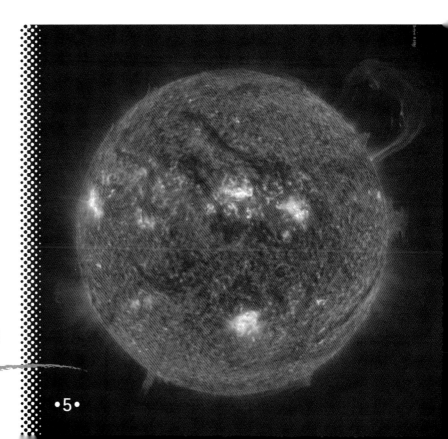

Where does the Sun's energy come from?

The Sun's energy comes from a place deep inside it, called the core. At the core, huge explosions release bursts of energy every second. Each burst of energy takes thousands of years to travel from the Sun's core to its surface. When the energy reaches the Sun's surface, it leaps into space and travels to the Earth, like waves crossing an ocean. Even though the Sun is 93 million miles (150 million km) from Earth, its energy reaches us in about eight minutes.

Fact file
Scientists believe that it will be about 10 billion years before the Sun begins to run out of energy.

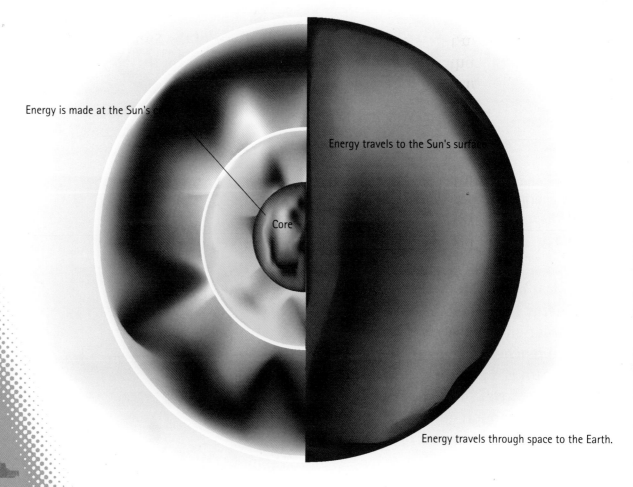

Energy is made at the Sun's core

Energy travels to the Sun's surface

Core

Energy travels through space to the Earth.

The Sun's energy travels through space and reaches the Earth as light and heat.

Using solar energy

Natural solar energy is used for growing **crops**, drying clothes, starting fires, and growing plants in **greenhouses**.

Natural solar energy cannot be switched on and off. The Sun does not shine at night and cloudy weather can block its energy. A source of energy that people could control became available in the 1800s. It was called electricity.

Electricity was made by burning coal and used to power factories and homes. People realized they had a source of energy to power their machines—a source that could be switched on and off, could be sent to wherever it was needed, and would never run out. They began to build bigger, more powerful machines that needed more electricity.

Today, the fossil fuels that are burned to make electricity are beginning to run low. It will be a long time, however, before solar energy runs out. Scientists have worked out ways to collect, store, and change the power of the Sun into electricity.

People use natural solar energy to dry the laundry.

Solar energy through history

People have been using the Sun's energy for thousands of years for light and heat.

Ancient use of solar energy

In ancient times, people used the Sun to dry fish and meat. This kept the food from **spoiling**. Animal skins were cleaned and dried in the Sun before being worn. Some houses were built from sun-dried mud bricks called adobe. Once built, adobe collects the Sun's heat during the day and releases it at night, keeping everyone warm inside.

The Sun was so important to ancient peoples that they believed it to be a god. If the Sun did not shine, people thought their god was angry and they became very afraid. They knew their crops would not grow without the Sun's help. Ancient people did not understand that the Sun was not a god. But they did understand they could not survive without it.

Fact file
The word "solar" comes from *sol*, an ancient Latin word that means "sun."

In some places, people still use the Sun's energy as they did in ancient times. Here, apricots are being dried in the Sun.

The machine age

Hundreds of years ago there were no machines. Animals did the heavy pulling and carrying work that people were not strong enough to do. As people needed more power, the first machines were invented. These machines were powered by water and wind. Water and wind were not reliable sources of energy. The wind did not always blow, and rivers froze during winter.

During the 1700s, steam-powered machines were invented which did the work of many people much faster. A lot of coal had to be collected and burned to make the steam. Bringing coal from out of the ground was very hard and dangerous work. Bringing large amounts of coal to the machines was difficult, because trains, trucks, and proper roads did not exist. Some scientists understood that the Sun was a source of energy and began to find out how that energy could be collected and used to power the hungry new machines.

Machines could do the work of many people much faster. Some machines were powered by water and steam.

Early solar technology

Scientists discovered that sunlight could be trapped and made more powerful. Using mirrors, sunlight was reflected onto specially shaped pieces of glass called lenses. Shining sunlight through a lens onto a small area beneath it could make the Sun's energy strong enough to boil water. In 1774, a French scientist called Antoine Lavoisier was able to make the sunlight so strong that it could be used to melt metal in a **furnace**.

Antoine Lavoisier's solar furnace, which he used in his chemistry experiments

In 1882, Augustin Mouchot and Abel Pifre invented a solar-powered printing press. The Sun's energy was captured in a dish-shaped mirror and reflected onto a boiler filled with water. The mirror made the Sun's energy strong enough to heat the water to steam, making the printing press work. The press printed 500 copies an hour of a newspaper called *The Sun Journal*.

Modern solar technology

Although early inventors found solar power very useful, it was not a very reliable source of energy because the Sun did not always shine. Inventors began to look for a new source of energy that was available whenever it was needed.

Electricity

During the 1800s, electricity became the new power that ran machines. Inventors began making lots of different machines that could use electricity. Today, there are millions and millions of machines in the world, all using electricity. The electricity is made from coal, oil, and gas, and these are beginning to run out.

The effects of burning fossil fuels to make electricity

Burning such huge amounts of fossil fuels to make electricity fills the air with tiny pieces of **soot**, dangerous gases, and chemicals. The soot can get into people's lungs and cause diseases such as **asthma** or **cancer**.

Some of the gases trap the Sun's heat, making the Earth hotter, which can upset the weather. Some of the chemicals mix with raindrops to make **acid** rain. When it rains, the acid kills trees and poisons lakes.

Scientists have now found ways to make electricity from the Sun's energy. Solar energy is safe, clean, and renewable.

These trees have been poisoned by acid rain.

Turning sunlight into electricity

One of the simplest technologies for turning sunlight into electricity is the solar cell.

Solar cells

A solar cell is made from layers of **silicon** coated with a special chemical. Sunlight falling onto a solar cell produces a small amount of electricity that is collected in metal wires on the cell's surface. One advantage of the solar cell is its small size. Three or four solar cells are enough to run a pocket calculator. However, several thousand are needed to power a solar car or a **satellite** in space.

Panels of solar cells are very useful in mountains or deserts, where it is too expensive or too difficult to build the towers that carry overhead electric wires. Solar cells can be used to run equipment in places where there is no one to look after it. The Sun shines everywhere, and panels of solar cells can be moved easily to wherever they are needed. The solar energy collected during daylight can be stored in **batteries** for use at night.

Fact file
The first solar cells were made in 1889 by Charles Fritts. The first solar panel was made by William J. Bailey in 1908.

Solar cells can be placed together on panels to power all sorts of equipment.

A solar cell turns sunlight into electricity.

Solar Power

Man cleaning solar panels in a village in Sudan, Africa

Solar power in poor countries

In some countries, many people are too poor to buy electricity or fuel to run **generators**. Solar panels can make enough electricity to run important equipment such as water pumps and telephones.

In Mali, Africa, there is no electricity or telephones and most people cannot read or write. Giving people warnings about new diseases or dirty drinking water is difficult. In 1993, a solar-powered radio station began broadcasting important information each day. The radio station has a small transmitter powered by 12 solar panels. Only three people are needed to operate the equipment, and everything is so light that it can be packed into a suitcase and loaded onto the back of a camel. The people of Mali listen to the broadcasts on small solar-powered or wind-up radios.

Storing solar energy

Solar cells cannot hold the electricity they make from sunlight. Wires connect the cells to batteries so the electricity can be stored for later use. Solar energy can also be collected and stored in pools of salty water called solar ponds.

Solar ponds

During the day, sunlight warms the water in the pond. The saltiest water soaks up the Sun's heat and sinks to the bottom, where it stays. The bottom of the pond has a black lining that **absorbs** the heat. The temperature here can be hot enough to **scald** the skin. Solar ponds can be covered to stop the heat from escaping. The hot water can be piped away to make electricity or used to supply homes with hot water. Tiny water animals are put into the water to eat any algae and keep the water clear so that plenty of sunlight can shine through.

Fact file
Some trees can store the solar energy captured by their leaves for hundreds of years.

Sunlight

Slightly salty cooler water

Very salty hot water

Black base absorbs heat

Cold water is piped in

Hot water is piped away

A solar pond collects and stores solar energy for later use.

Solar Power

Solar farms

When larger amounts of electricity are needed, solar farms are used. Solar farms have rows of solar cells on large, flat panels that are tilted towards the Sun. The electricity produced on a solar farm is not stored in batteries. It is passed through a special machine that changes it into the type of electricity we use in our homes. It is then sent through overhead powerlines to nearby towns and cities.

Solar farms depend on sunshine and can only make electricity during the day. A lot of flat, open space is needed to house the large number of solar panels on each solar farm. Countries such as Australia and the United States have plenty of space and sunshine, but solar farms are not suitable for all parts of the world.

This solar farm has nearly 7,000 solar panels. It is as big as five football fields and provides enough electricity for 6,000 homes.

Some solar power stations use rows of vacuum-tube collectors to capture the Sun's energy.

Solar power stations

Electricity can also be made by solar power stations. A power station changes energy into electricity by using blasts of steam to spin giant wheels called turbines. A generator turns this spinning movement into electricity. In a power station, fossil fuels are burned to heat water to make the steam. In a solar power station, the Sun's heat is the energy used to heat the water to make the steam. The solar energy is collected by using different types of mirrors.

Some solar power stations use rows of curved mirrors called **vacuum**-tube collectors. A glass tube runs through the center of each row of mirrors. Inside this tube is a copper pipe filled with a liquid. The mirrors bounce sunlight onto the glass tube, heating the liquid inside the copper pipe. All the air inside the glass tube is sucked out. This stops heat escaping from the liquid inside the pipe. The hot liquid is piped away to make steam for electricity.

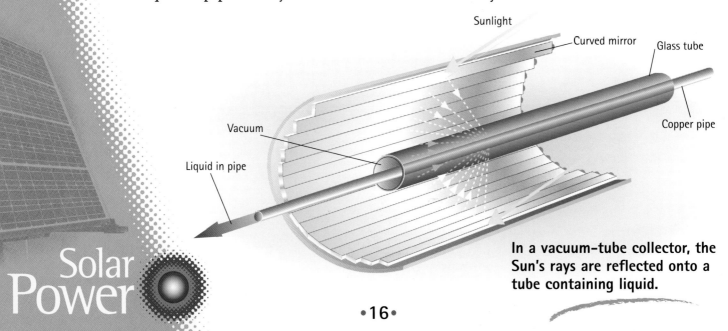

In a vacuum-tube collector, the Sun's rays are reflected onto a tube containing liquid.

Power towers

A power tower is a type of solar power station. Power towers are surrounded by thousands of flat mirrors. When the mirrors reflect sunlight onto a container of liquid at the top of the tower, the liquid becomes very hot. It is then piped away to heat water to steam. The steam is used to power generators that make electricity.

Solar chimney

A solar chimney is another type of power station. A solar chimney uses hot air instead of hot liquid to make electricity. The chimney is surrounded by a **canopy** of special glass that allows the Sun's energy to heat the ground and air beneath it.

The container glows white-hot from the heat reflected by the mirrors. Inside the container, a liquid is heated and piped away to make steam.

Mirrors reflect sunlight onto the tower.

Solar Two is a solar power tower built in the U.S. in 1996. The heat reflected by its mirrors is the same as 600 suns.

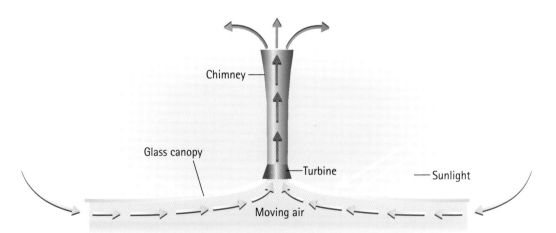

If this solar chimney is built, it will be the tallest human-made structure in the world. It will be seen 62 miles (100 km) away.

As the hot air moves under the canopy towards the tower, it spins the turbines placed around the tower's base. The turbines drive a generator that produces electricity. At night, the heat that has been trapped in the ground will rise and move towards the tower. Even in darkness, the generator will still be able to make some electricity.

Solar power at work

Around the world, people are becoming more energy conscious. They know that safe, clean energy sources are available. More people are using solar technology as it improves and becomes cheaper.

Greenhouses

A greenhouse uses solar energy for growing plants. The glass it is made from **intensifies** the Sun's rays. The plants get so much solar energy that they grow more quickly. The glass canopy that surrounds a solar chimney is like a greenhouse. Plants have been grown beneath solar chimneys, too.

A greenhouse uses natural solar energy to keep plants healthy and warm.

Houses

Houses can be designed to make the best use of natural solar energy—even in very cold countries, where there is less daylight in winter. By facing a house in a certain direction and using the right building materials, less electricity is needed for heating and cooling.

Architects (*AR-ki-tekts*) use special computer programs to help them design buildings that are more **energy-efficient**. Some governments offer prizes to encourage as many energy-efficient designs as possible.

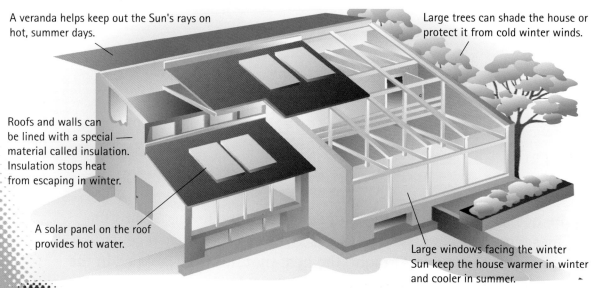

Houses can be designed to make the best use of natural solar energy.

Solar water heaters

A solar water heater is a special type of rooftop solar panel that uses natural solar energy to heat water. The panel is a thin, glass-covered box, with its inside painted dull black. Water flows slowly through a narrow pipe inside the box. The pipe is painted black, too. This is because black soaks up the Sun's heat more easily than any other color. The water absorbs the Sun's heat and is stored in a tank until it is needed.

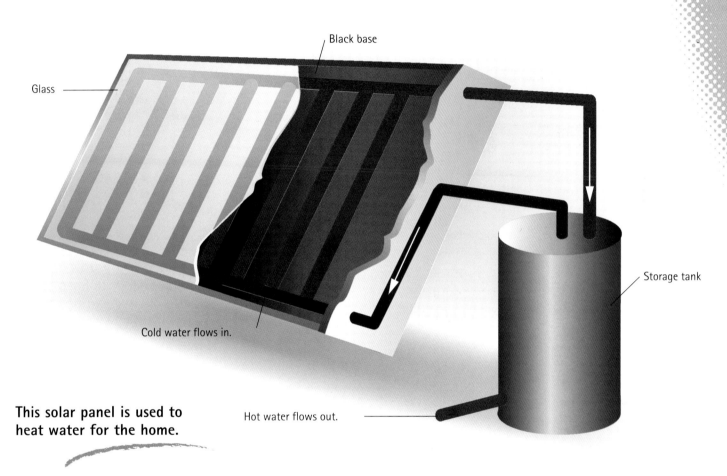

This solar panel is used to heat water for the home.

Swimming pools

Swimming pools can be heated by solar energy, too. Black plastic or rubber tubing is fixed to the roof on the side of the house that gets the most sunlight. Water from the pool is pumped through the tubing. The more tubing on the roof, the warmer the water becomes, because it takes longer to flow through it. A pool can also be covered by a special blanket that traps the solar energy beneath it.

Fact file

The first solar water heater was built in 1767 by a Swiss scientist named Horace Benedict de Saussure. It was a simple wooden box with a glass cover and a black base. When it was placed in the Sun, it made the water very hot.

An experimental water purifier floats on a moat in Tokyo, Japan. Its solar panels generate electricity and purify the water with that energy.

Fact file

In 1874, the world's largest solar still was built in Chile, South America. It could produce 6,000 gallons (22,700 l) of drinking water every day. It did so for 40 years.

Solar stills

A solar **still** is used to turn salty seawater into drinking water. Seawater is pumped into closed glass containers beneath the ground. As the Sun warms the glass, it slowly turns the seawater into steam, leaving behind crystals of salt. When the steam touches a cooler part of the walls or roof, it changes back into droplets of **freshwater**. The droplets run down into a deep tray at the bottom.

Solar stills are used in parts of the world that are close to the sea and have no rivers or streams that provide fresh drinking water.

Solar Power

Solar furnaces

A solar furnace is used to heat metals to very high temperatures. A solar furnace has lots of mirrors that collect sunlight. Some of the mirrors have motors inside them. The motors move the mirrors so that they follow the Sun as it passes across the sky. A steering system tilts the mirrors so that they catch as much sunlight as possible. Mirrors that follow the Sun like this are called heliostats.

The heliostats reflect the sunlight onto **fixed** mirrors. The fixed mirrors **concentrate** the sunlight they receive into a very small space on the tower, called a target. This makes the solar energy thousands of times more powerful.

Fact file

- The Odeillo solar furnace, in France, is the largest one in the world.
- Each of its 63 heliostats is 24 feet (7.3 m) high and 20 feet (6.1 m) across.
- Its giant fixed mirror is 165 feet (50 m) high and 137 feet (42 m) across.
- It produces temperatures almost half as hot as the surface of the Sun itself.
- This solar furnace makes the Sun's energy up to 12,000 times more powerful.
- By opening and closing special shutters in front of the target, the furnace can be turned on and off in less than a second. The shutters have to be cooled to stop them from melting or shattering as the temperature suddenly changes.
- French scientists use the Odeillo solar furnace to test materials and metals that will be used in spacecraft.

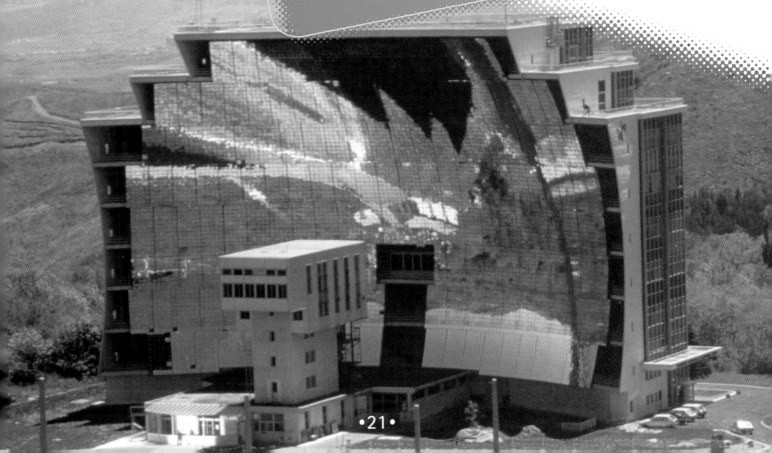

The solar furnace at Odeillo, in France

Solar electricity

Some people use rooftop solar panels to make electricity for their homes from sunlight. The panels provide electricity during the day to run **appliances**. At night, the main community electricity supply is switched on so that appliances keep running and lights can be used.

Around the world, governments are encouraging people to use solar energy in their homes by helping them with the cost of solar water heaters and electricity systems. The more people use solar energy, the less pollution there will be and the longer fossil fuels will last.

Appliances

Most homes have electrical appliances such as refrigerators, microwave ovens, washing machines, televisions, and computers. Energy experts have given appliances special energy ratings. If an appliance has five stars, it means it is very energy-efficient and uses less electricity. Low-energy lightbulbs last about eight times as long as normal lightbulbs and use only one-quarter of the electricity.

Turning off lights and appliances that are not being used saves electricity. Saving electricity means burning less fossil fuels. This is kinder to the environment and people's health.

> ### Fact file
> A family that used solar panels to make electricity and heat water for two years saved 4.7 tons (4.3 t) of coal from being burned and prevented 9.1 tons (8.3 t) of greenhouse gases from polluting the air.

Low-energy lightbulbs use less electricity and last longer than normal lightbulbs.

Solar-powered transportation

Most of our transportation uses up precious fossil fuels and pollutes the air. People are trying solar-powered transportation to save **gasoline** and keep the air cleaner and safer for everyone to breathe.

Solar-powered cars

A solar-powered car looks very different to a normal car. Solar-powered cars are made as light as possible because the electric motors used to power them are very small. The cars have smooth edges and are close to the ground, so they can slip through the air easily. Solar cars do not produce any poisonous gases and chemicals, but are very expensive to build and cannot go very fast. The heavy batteries they carry can slow them down. The batteries are needed to store electricity for when there is no sunshine.

Solar-powered cars will probably never be practical enough for everyday use. They have helped scientists to invent better solar cells that can be used for other solar-powered equipment.

This solar-powered car won the 2001 World Solar Challenge in Australia. *Nuna* was powered by solar cells that were as thin as paper. Two strips of solar cells came from the *Hubble Space Telescope*.

This electric truck is recharging its batteries using solar energy from the solar panels above.

This solar-powered vehicle, in Germany, is being developed as an alternative to fossil-fueled transportation.

Solar rickshaws

Hand-pulled carts called rickshaws were once used in Asia. Nowadays, the rickshaws have motors that use fossil fuels and add to the choking pollution of Asia's overcrowded cities. A British **environmentalist** named Malcolm Moss invented a solar-powered rickshaw and drove it around India in 2001. The rickshaw reached speeds of 25 miles (40 km) per hour. A small electric motor gave the rickshaw the extra power it needed to pull away from traffic lights.

The rickshaw was shown on five television channels. One channel had an audience of 10 million people! The Indian government is excited about the rickshaw and is helping everyone to understand that it will keep the cities cleaner and protect people's health. If 10,000 rickshaws run for 10 years, they will have prevented more than a million tons of greenhouse gases from escaping into the air.

Solar Power

Solar boats

In Australia, in February 2000, Robert Dane launched the world's first wind- and solar-powered ferry, *Sydney Solar Sailor*. It is the world's largest solar-powered sea vessel and carries 100 passengers.

Solar power is perfect for boats. Boats do not need extra power to travel uphill or to travel with heavy traffic. There are also no buildings to block the sunlight. The solar sails capture light straight from the Sun as well as sunlight reflected off the water. The weight of the storage batteries helps to keep bigger boats afloat.

Sydney Solar Sailor is the first wind- and solar-powered ferry in the world. It has solar panels instead of sails.

Fact file

In 1999, Japanese sailor Kenichi Horie traveled 10,000 miles (16,000 km) from Ecuador, in South America, to Japan in his solar-powered boat, *Malt's Mermaid II*. It was made from 22,000 recycled cans.

Helios is a solar-powered aircraft. The solar cells on its 242-foot (74-m) wing produce enough electricity to power 100 homes.

Solar-powered aircraft

Solar-powered aircraft use solar energy to turn the propellers on their wings. These aircraft are only experiments and do not carry pilots. *Helios* is an experimental solar-powered aircraft that looks like a giant boomerang. The aircraft's 14 propellers are powered by 62,000 solar cells. With wings longer than a Boeing 747, *Helios* weighs less than most cars. In July 2001, *Helios* flew as high as 75,000 feet (22,800 m) on its first test flight.

It is hoped that solar-powered aircraft will be able to take over for some satellites. It cost $15 million to build *Helios*, but that is much cheaper than building and launching a satellite.

Fact file

The world's first solar-powered aircraft flew 163 miles (260 km) from Paris across the English Channel in 1981. *Solar Challenger* flew as high as 11,000 feet (3,350 m). It was powered by 16,000 solar cells attached to its wings and tail.

The *Hubble Space Telescope* is the largest satellite to have been sent into space.

Satellites

Solar energy has been used to power spacecraft since 1958. Satellites are small spacecraft sent into space attached to rockets. Once in space, their solar panels are turned to face the Sun. Solar energy keeps them in **orbit** and powers the cameras and computers inside them.

Solar sails

A solar sail is a spacecraft made of a special material as thin as a garbage bag. Its eight triangular blades make it look like a strange windmill. Like a satellite, a solar sail needs a rocket to carry it into space. Once in space, the sail uses solar energy to move it along. A solar sail does not use solar energy in the same way that a satellite does. A solar sail is not covered with solar cells because it does not change sunlight into electricity. The sunlight bounces off the sail and this pushes it along through space. On July 25, 2001, a rocket carrying a solar sail was launched from a submarine near Russia. Unfortunately, the rocket crashed.

A remote-controlled, solar-powered robot like *Sojourner* can explore planets, collect rocks and soil, and send back pictures for scientists to study.

Solar power in the future

Solar energy is becoming a more reliable source of safe, clean, renewable energy. Using solar energy is a good way to save valuable fossil fuels and protect the environment and people's health. Solar technology is getting better all the time and becoming cheaper to make. As this happens, more people will use it in their everyday lives. Solar energy is particularly suitable for people living in **remote** places.

Space colonies

In the future, humans may have to live in faraway places such as the Moon or other planets. To learn how to do this, scientists built Biosphere II in the U.S. A biosphere is like a mini Earth inside a gigantic greenhouse. It has plants, trees, animals, and insects. Biosphere II even had a rain forest, a desert, and an ocean filled with more than 1 million gallons (3 million l) of salt water.

In 1991, eight people spent two years living in Biosphere II. All the air, water, and food was recycled inside the structure. Heating and lighting were powered by solar energy. A thousand **sensors** sent information to computers and scientists so they could understand how the different environments worked.

Biosphere II contained 3,800 types of plants and animals.

One day, people on Earth may get their electricity from a solar power station in space, like this one.

Solar power stations in space

In the future, it could be easier to make electricity in space. Solar cells work better without the Earth's atmosphere blocking the Sun's rays. Huge solar panels in space could produce electricity that would be changed into invisible waves called microwaves. Special receiving stations on Earth would change the microwaves back into electricity and send it to people's homes. Much care would be needed. If the microwaves missed their target they would "cook" whatever they touched!

Solar power from deserts

Hot deserts seem like perfect places for collecting solar energy. Deserts get very little or no rain, so the Sun's energy is available every day. Deserts cover one-fifth of the Earth's surface and few people live in them. There would be plenty of space for solar chimneys or power towers in the world's deserts. Unfortunately, deserts have strong winds that blow the sand about. If sand settles on the solar panels or mirrors it would block out some of the sunlight. Their smooth surfaces would be scratched as sand moved across them.

If scientists can find a way to protect the solar panels and mirrors from the sand, the world's deserts might become a future source of solar power.

Advantages and disadvantages of solar power

Some people think that solar technology is ugly.

Fossil fuels are a non-renewable source of energy. If we keep using them at the current rate,
- coal will run out in 250 years
- oil will run out in 90 years
- gas will run out in 60 years.

There are other sources of energy that are cleaner, safer, and will not run out. Solar energy is a safe, clean, and renewable source of energy for the world's future power needs.

ADVANTAGES OF SOLAR POWER	DISADVANTAGES OF SOLAR POWER
• Solar energy does not cause pollution.	• Darkness at night and cloudy weather reduces the sunlight available.
• Solar energy is useful for remote places. It can be tailored for use in individual households (using, for example, solar cells, solar panels, and solar water heaters) as well as for larger areas.	• A lot of space is needed for solar power stations.
	• Some solar technology is still expensive to make.
• Sunlight is free and will not run out.	• Fossil-fuel energy is used to make some of the parts for producing solar energy (such as solar cells), and a small amount of pollution is produced.

Glossary

absorbs soaks up

acid a type of chemical that can be harmful to people and the environment

appliances machines that are designed to do a particular job

architects people who design buildings

asthma a disease which makes breathing difficult

batteries containers filled with chemicals that can store or produce electricity

cancer a serious disease where damaged cells grow into lumps called tumors that stop the body from working properly

canopy an overhead cover

concentrate to squeeze a lot into a small space

crops plants grown for food

energy-efficient using energy without waste

environmentalist someone who cares for the environment

fixed not moving

freshwater water that is not salty

furnace a container for a large, very hot fire

gasoline a liquid fuel made from oil that is burned inside an engine

generators machines that turn energy into electricity

greenhouses glass buildings in which plants are grown or kept warm

Hubble Space Telescope an American space telescope that was launched into space on a satellite in 1990

intensifies makes something stronger

orbit to go around an object

remote very far away from other people

satellite a spacecraft that circles the Earth and sends and receives information

scald to burn with very hot liquid

sensors instruments that react to change

silicon a substance found in sand and some rocks

soot a black powder that rises with the smoke when coal is burned

spoiling going bad and becoming unfit to be eaten

still equipment that heats liquid to steam and cools the steam, turning it back into liquid

vacuum a space which has no air in it

Index

A
aircraft   26
architecture   18

B
boats   25

C
cars   23

E
Earth   5, 6
electricity   7, 11, 12–13, 14–17, 22

F
food chain   5
fossil fuels   4, 7, 9, 11, 30
furnaces, solar   10, 21

G
greenhouse effect   11
greenhouses   18

H
heliostats   21
houses   8, 18

L
Lavoisier, Antoine   10

M
machine age   9

N
non-renewable energy   4, 30

O
Odeillo solar furnace   21

P
pollution   11, 22, 30
power stations, solar   16–17, 29, 30
power towers   17

R
renewable energy   4, 5, 28, 30
rickshaws   24
robots   27

S
satellites   27
solar cells   12, 15, 30
solar chimneys   17, 18
solar farms   15, 29
solar ponds   14
solar stills   20
solar technology   10–13, 16–17, 19–27, 28
space colonies   28
spacecraft   27
steam power   9
storing energy   14
Sun   5, 6

T
transportation   23–7

W
water heaters   14, 19, 30